Chasing the LOCH NESS MONSTER

CHASING CRYPTIDS

Anna Anderhagen

Big Buddy Books
An Imprint of Abdo Publishing
abdobooks.com

abdobooks.com

Published by Abdo Publishing, a division of ABDO, PO Box 398166, Minneapolis, Minnesota 55439.

Printed in the United States of America, North Mankato, Minnesota
102023
012024

THIS BOOK CONTAINS RECYCLED MATERIALS

Design: Denise Hamernik, Mighty Media, Inc.
Production: Mighty Media, Inc.
Editor: Liz Salzmann
Cover Photograph: primopiano/Adobe Stock
Interior Photographs: Alizada Studios/Shutterstock Images, p. 7; Daniel Eskridge/Adobe Stock, pp. 13, 23; Elzbieta Sekowska/Shutterstock Images, p. 21; Khadi Ganiev/Shutterstock Images, p. 5; Kit Leong/Shutterstock Images, p. 25; Matt Ragen/Shutterstock Images, p. 11; Michael Rosskothen/Shutterstock Images, p. 15; mikefergusonphotography/Shutterstock Images, p. 9; Orest lyzhechka/Shutterstock Images, p. 29; PA Images/Alamy Photo, p. 19; Steve Challice/Cover Images/AP Images, p. 17
Design Elements: Flavijus Piliponis/Adobe Stock (silhouette); Joko/Adobe Stock (tape and paper); Melica/Adobe Stock (polaroid frame); Net Vector/Shutterstock Images (map); Oleg Iatsun/Shutterstock Images (compass); Piman Khrutmuang/Adobe Stock (paper); STILLFX/Shutterstock Images (background texture); TatianaMakhakhei/Shutterstock Images (footprints); Wandeaw/Shutterstock Images (background texture); Yevhenii/Adobe Stock (tape)

Library of Congress Control Number: 2023939278

Publisher's Cataloging-in-Publication Data
Names: Anderhagen, Anna, author.
Title: Chasing the Loch Ness monster / by Anna Anderhagen
Description: Minneapolis, Minnesota : Abdo Publishing, 2024 | Series: Chasing cryptids | Includes online resources and index.
Identifiers: ISBN 9781098291914 (lib. bdg.) | ISBN 9781098278816 (ebook)
Subjects: LCSH: Loch Ness monster--Juvenile literature. | Monsters--Juvenile literature. | Animals, Mythical--Juvenile literature. | Folklore--Juvenile literature. | Cryptozoology--Juvenile literature.
Classification: DDC 001.944--dc23

CONTENTS

CHAPTER 1

CHASING CRYPTIDS

You are fishing in a boat. Suddenly, you see a large fin in the water. You also see two huge humps! A beast stretches its long neck and head out of the water. It looks at you before diving back underwater.

Could this be the Loch Ness Monster cryptid you have been looking for?

FAST FACT

There have been more than 1,100 recorded Loch Ness Monster sightings.

The Loch Ness Monster is also known by the nickname “Nessie.”

CHAPTER 2

WHAT IS A CRYPTID?

A cryptid is an animal that has not been proven to exist. One of the best-known cryptids is the Loch Ness Monster. It lives in Loch Ness, a lake in Scotland. Many people think they have seen the Loch Ness Monster. But no one has been able to prove it. Most people **describe** the Loch Ness Monster as having a long neck, fins, and humps.

There are several Loch Ness Monster statues in towns and ponds near the lake.

CHAPTER 3

LOCH NESS

Loch Ness is located near the city of Inverness, Scotland. The lake is 23 miles (37 km) long. It is 788 feet (240 m) deep. Because the lake is so deep, the water is very cold.

The earliest stories of the Loch Ness Monster are from the Pict people. They lived in Scotland from about 300 to 900 CE. The Picts made stone carvings of a large beast with flippers.

Loch means "lake" in the Scottish Gaelic language.

CHAPTER 4

FIRST RECORDED SIGHTING

The first written account of the Loch Ness Monster is a story about St. Columba in 565 CE. St. Columba was a **monk** from Ireland. He saw a giant underwater monster bite a swimmer in Loch Ness. The beast was about to attack another man. But St. Columba yelled, “Go back!” The **colossal** beast obeyed him.

St. Columba is known for spreading Christianity throughout Scotland and Ireland.

CHAPTER 5

MORE SIGHTINGS

In 1933, two people were driving near Loch Ness. They saw a giant monster playing in the water near Abriachan Pier. The monster looked like a whale jumping and rolling on the water's surface. Soon, the creature dove back into the water with a loud splash.

CRYPTID PROFILE

NAMES: Loch Ness Monster, Nessie

CLASSIFICATION: marine animal

COUNTRY: Scotland

HABITAT: water

DESCRIPTION:
- long neck
- fins
- humps

PROVEN TO EXIST: not yet

In 1955, fourteen-year-old Ronald Mackintosh was on a drive near Loch Ness with his mother and brother. He looked out the window toward the lake. He shouted, “Stop the car!”

The family saw a **commotion** in the middle of the lake. A gray-brown beast **emerged** from the water. It was the size of a bus! It flipped over and crashed down. The monster made waves three feet (0.9 m) high.

People believe the Loch Ness Monster could be up to 30 feet (9.1 m) long.

CHAPTER 6

SHIMMERING CREATURE

In 2022, Jamie Huntley and Warren Speed were driving past Loch Ness. Huntley saw something sticking about seven feet (2.1 m) out of the water. It had a hump with gray, black, and brown patches on it. The creature **shimmered** in the water like a fish.

In 2019, tourist Steve Challice took this photo of a mysterious creature in Loch Ness. He thinks it's a giant fish. But others believe it is the Loch Ness Monster!

CHAPTER 7

TRACKING THE LOCH NESS MONSTER

Inventor Robert Rines tracked the Loch Ness Monster for more than 25 years. In 1972, he took an underwater picture of a large flipper in Loch Ness. He took another photo showing the monster's outline and size. But Rines could not prove that his pictures were of the monster. Some **researchers** think they showed objects in the lake.

Robert Rines (*right*) with one of his Loch Ness Monster pictures. Rines used an underwater camera to take pictures every 45 seconds.

CHAPTER 8

SONAR SEARCH

Loch Ness has very dark water. It is hard to see underwater, especially in deep areas. Sonar can help find things underwater using sound waves. In 1987, twenty boats did a sonar search in Loch Ness to look for the monster. The search found three large **unidentified** underwater objects, but no monster.

Sonar sends sound waves under the water. The sound waves echo back if they hit something.

CHAPTER 9

GIANT EELS IN LOCH NESS?

In 2018, scientists collected water samples from Loch Ness. They took **DNA** samples to see what animals lived there. Scientists **identified** DNA from deer, birds, and other common animals. They also found that thousands of eels live in Loch Ness. Some people believe the Loch Ness Monster is a giant eel.

Some people believe the Loch Ness Monster is a descendant of the extinct reptile plesiosaur (*pictured*). But the 2018 DNA study found no plesiosaur DNA.

CHAPTER 10

CRYPTID KEEPER: GARY CAMPBELL

Gary Campbell lives in Inverness. He runs the official Loch Ness Monster sightings **register**. The register includes more than 1,145 sightings dating back to 565 CE.

In 1996, Campbell saw the Loch Ness Monster. He saw a black hump coming out of the water. He thought it looked like a small whale.

The Loch Ness Exhibition Centre in Drumnadrochit is a popular tourist attraction. It features scientific research about the Loch Ness Monster.

NESSIE SIGHTINGS

Many sightings of the Loch Ness Monster happened near Aldourie Castle. Other places include Loch Ness's north side, Boleskine Cemetery, Abriachan Pier, and Drumnadrochit.

CHAPTER 11

DOES IT EXIST?

Thousands of people think they have seen the Loch Ness Monster. Some even claim to have taken photos of it. But no one can prove that the monster is real.

Most scientists do not think the Loch Ness Monster exists. They suggest that people are mistaking swimming deer or large eels for the Loch Ness Monster. What do you think?

Floating logs can look like the Loch Ness Monster's long neck.

GLOSSARY

colossal—huge, giant.

commotion—an excited disturbance.

describe—to tell about something with words. Such a telling is a description.

DNA—deoxyribonucleic acid. A material in plants and animals that carries genetic information.

emerge—to come out into view.

identify—to find out what something is.

monk (MUHNGK)—a man who is part of a religious order and lives in a monastery.

register—an official list of records.

researcher—a person who carefully studies a subject in order to learn facts about it.

shimmer—to shine with light that seems to waver slightly.

unidentified—not known.

ONLINE RESOURCES

Booklinks
NONFICTION NETWORK
FREE! ONLINE NONFICTION RESOURCES

To learn more about the Loch Ness Monster, please visit **abdobooklinks.com** or scan this QR code. These links are routinely monitored and updated to provide the most current information available.

INDEX